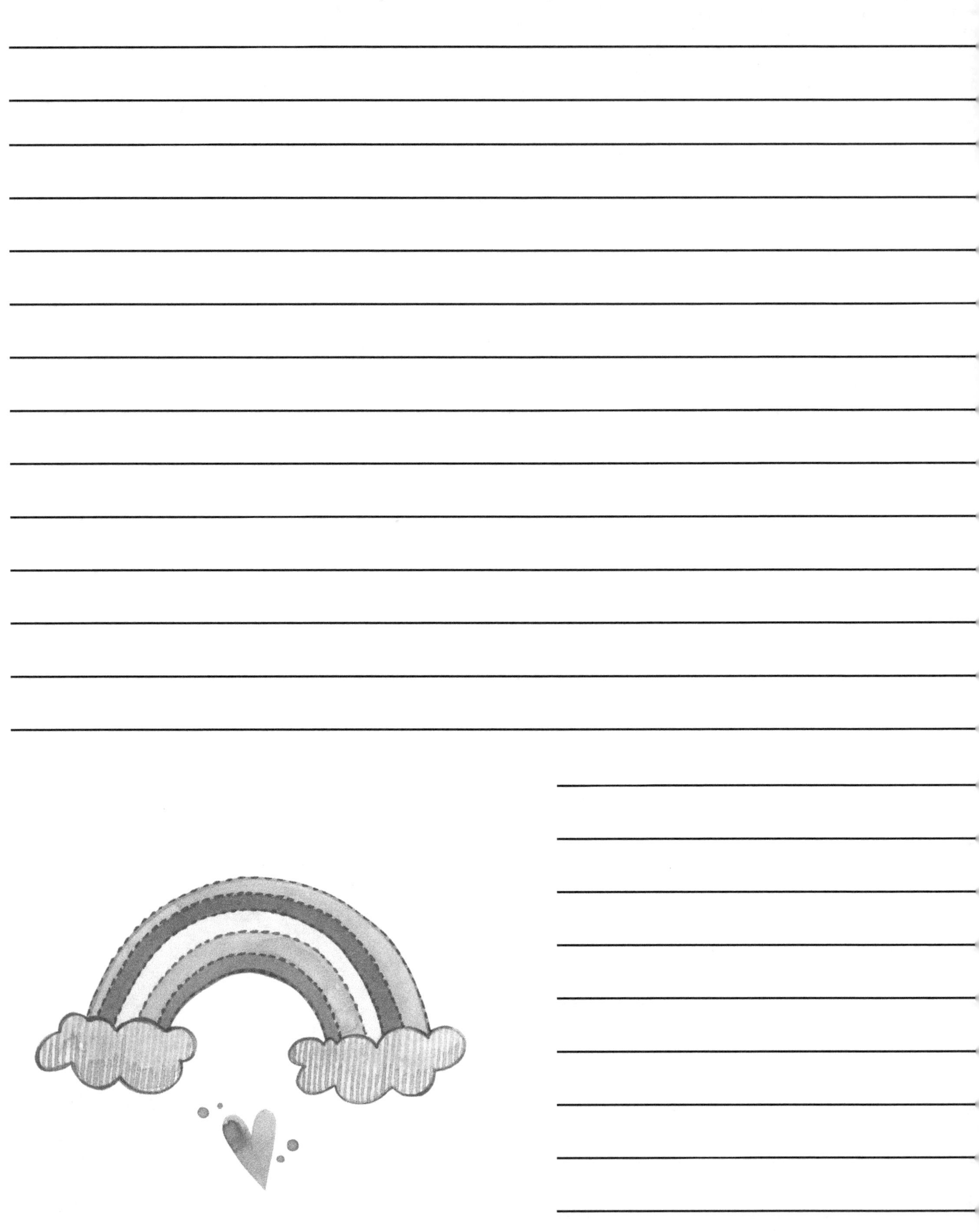

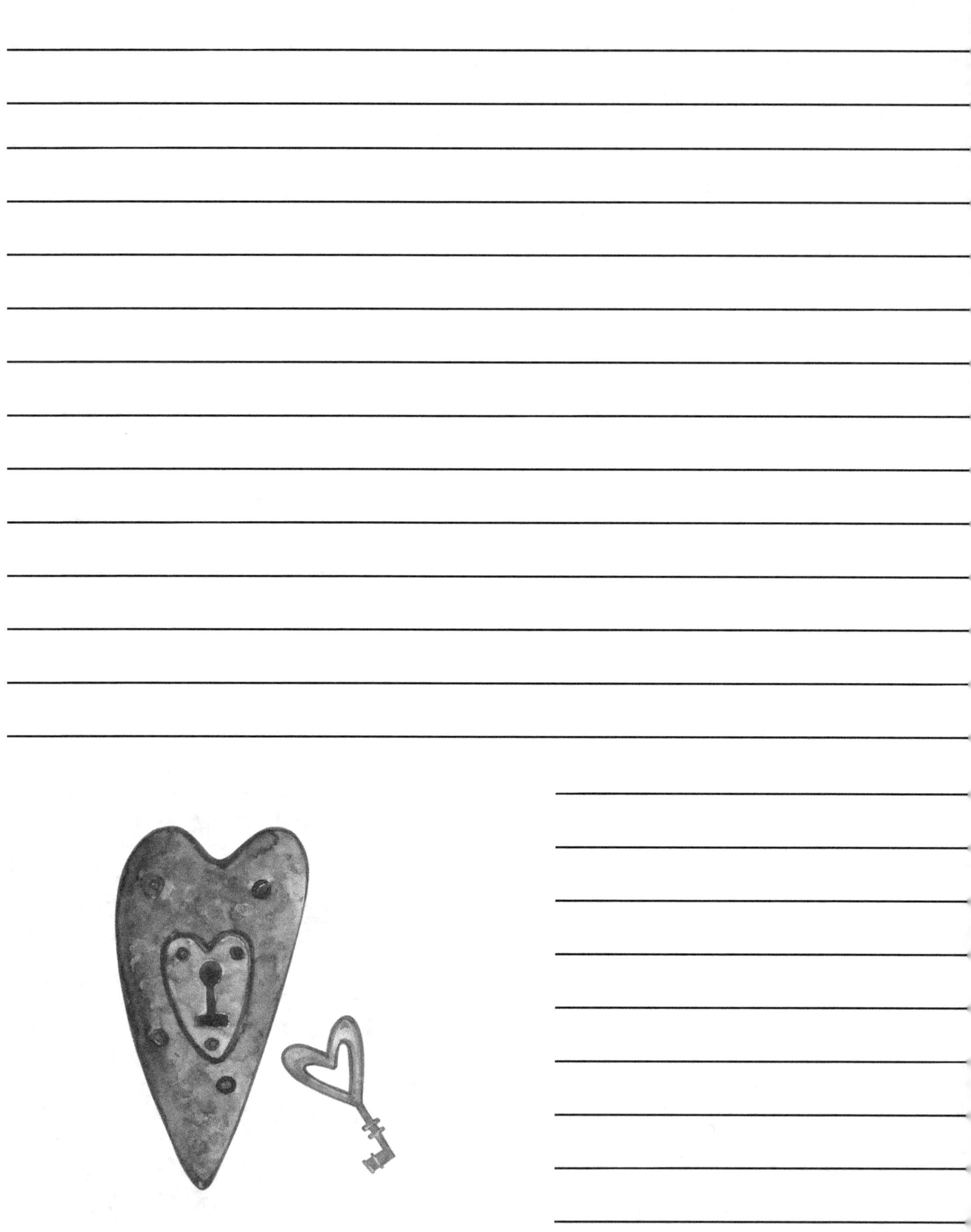

you + me = love

i love
you

love you

you are loved

love
is all
you
need

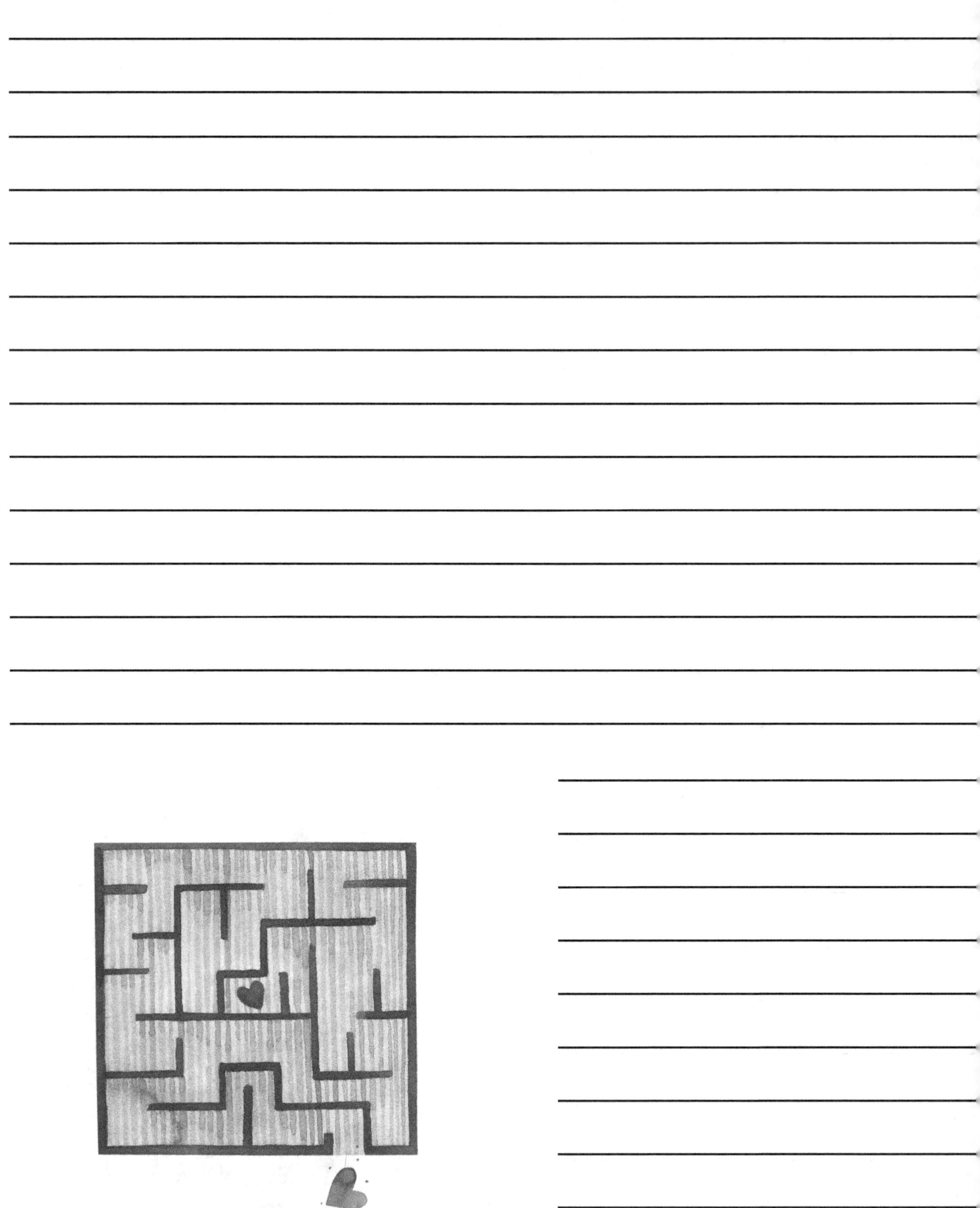

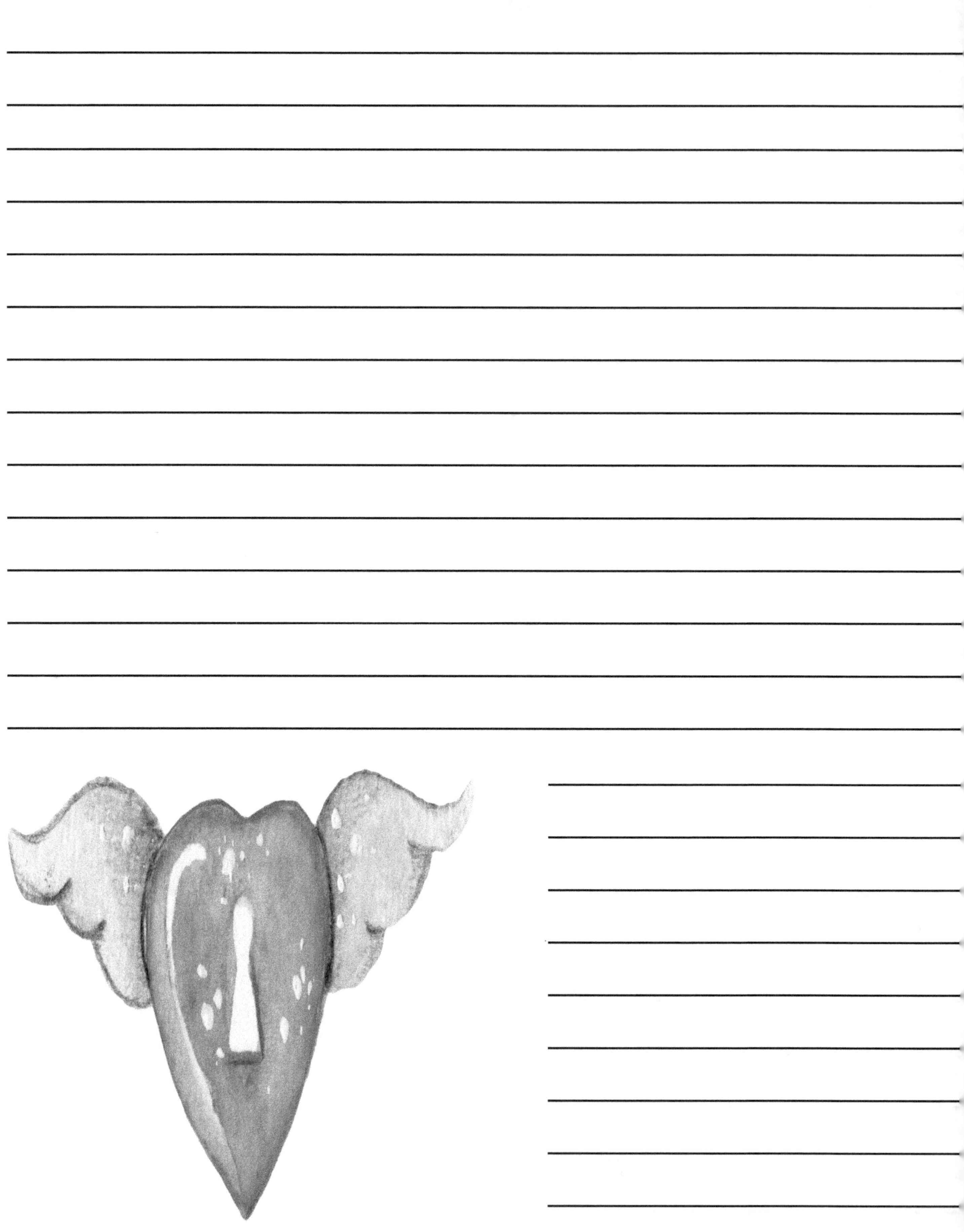

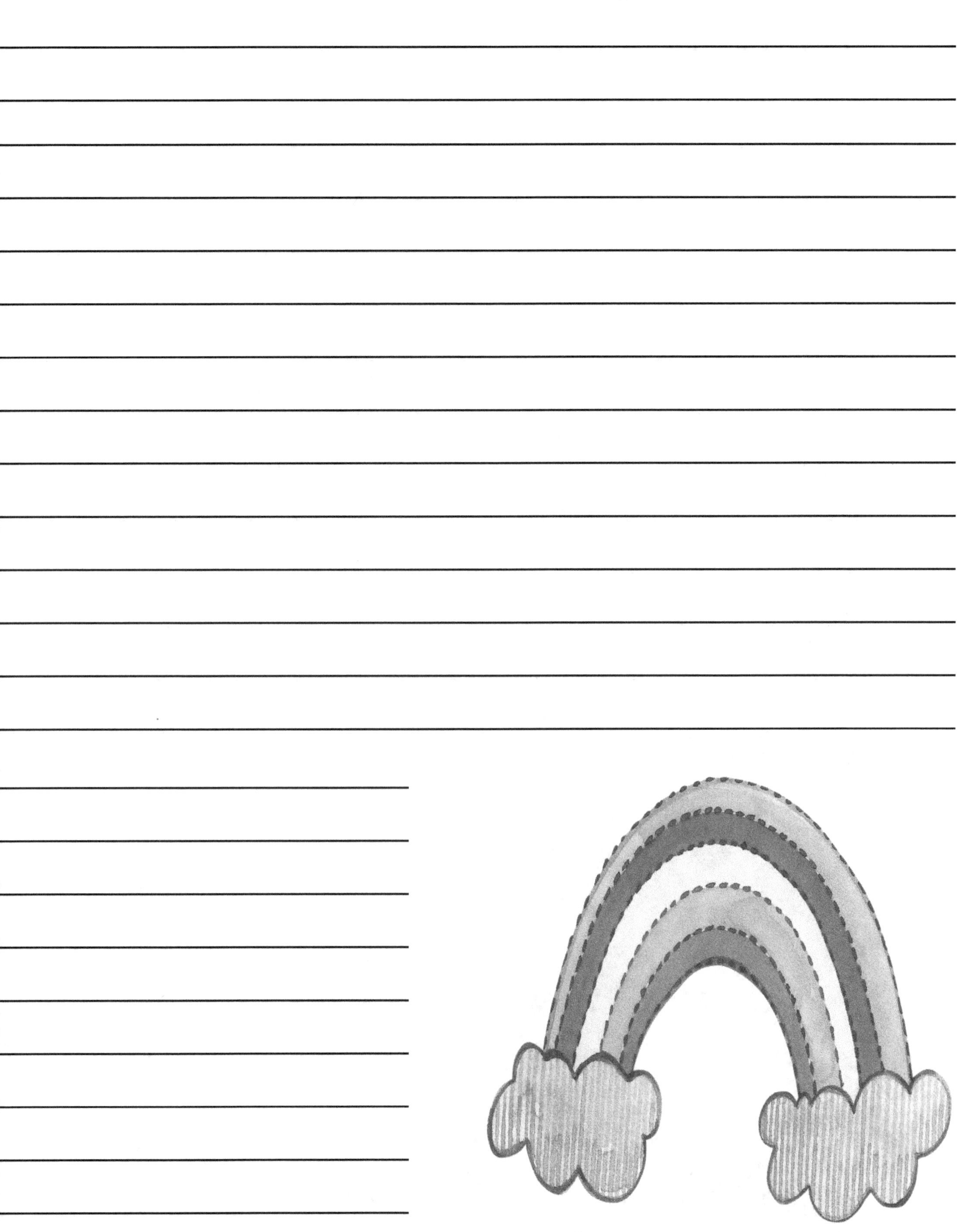

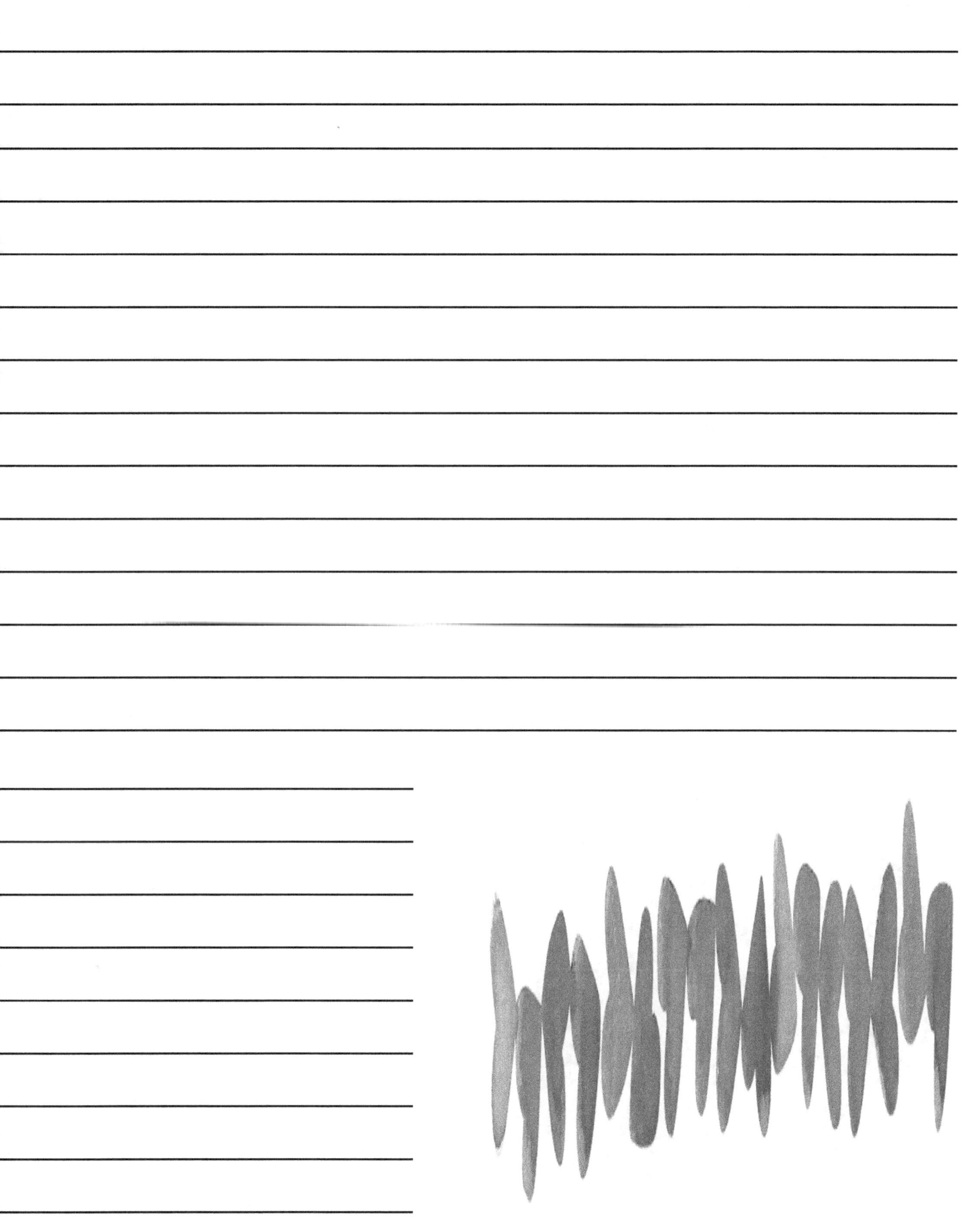

you
+
me
=
love

believe
IN
Love

i love
you

love
you

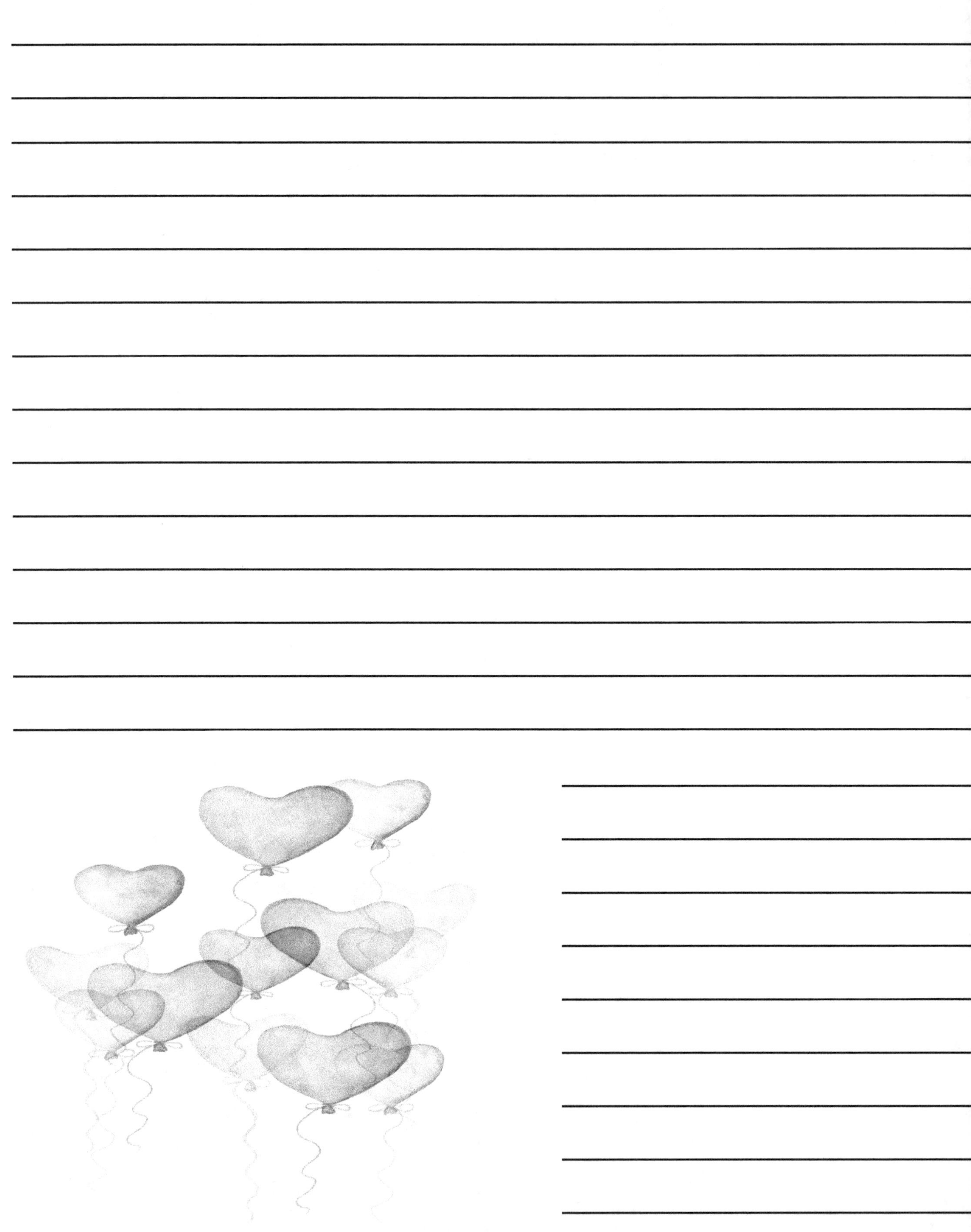

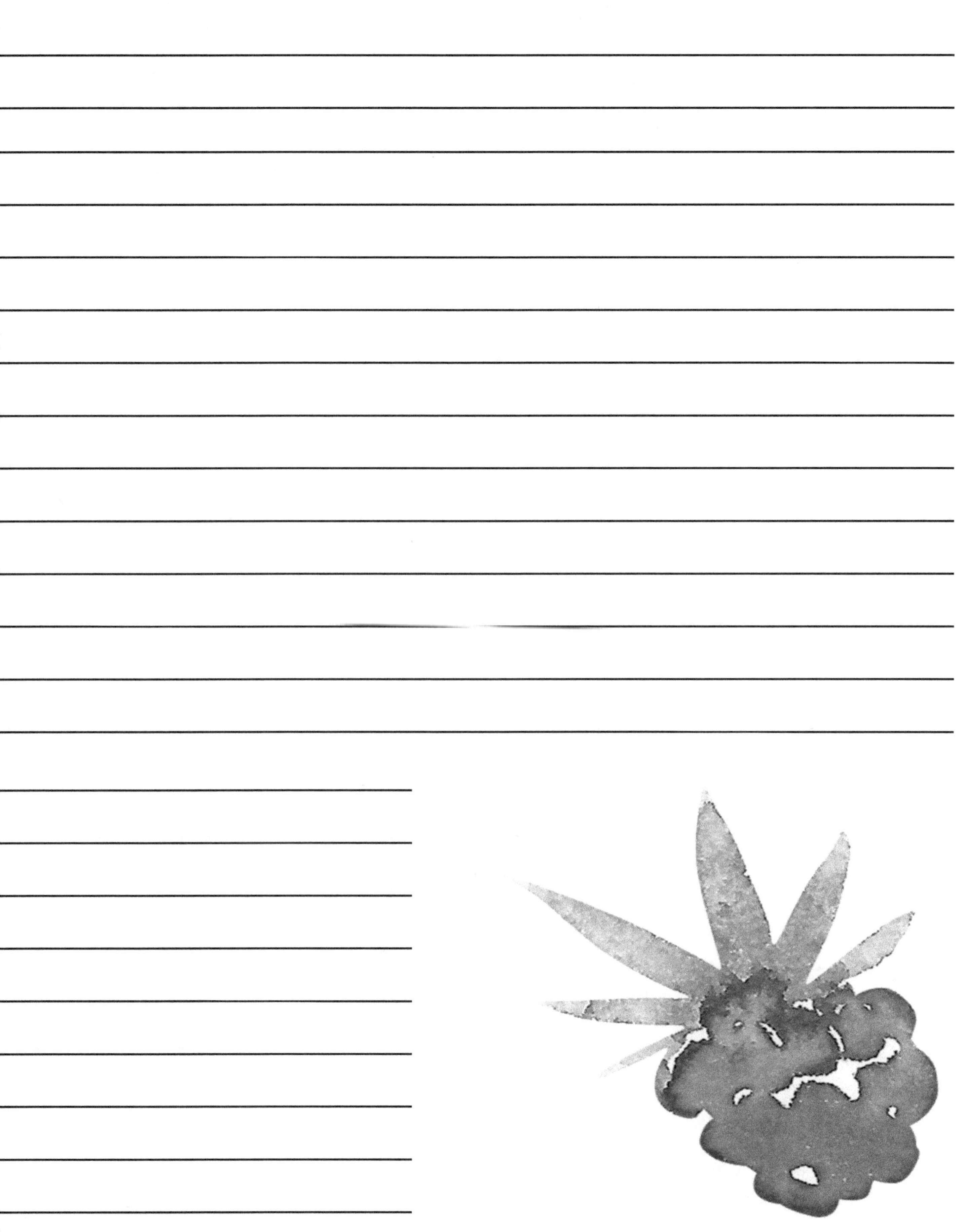

you are loved

love
is all
you
need

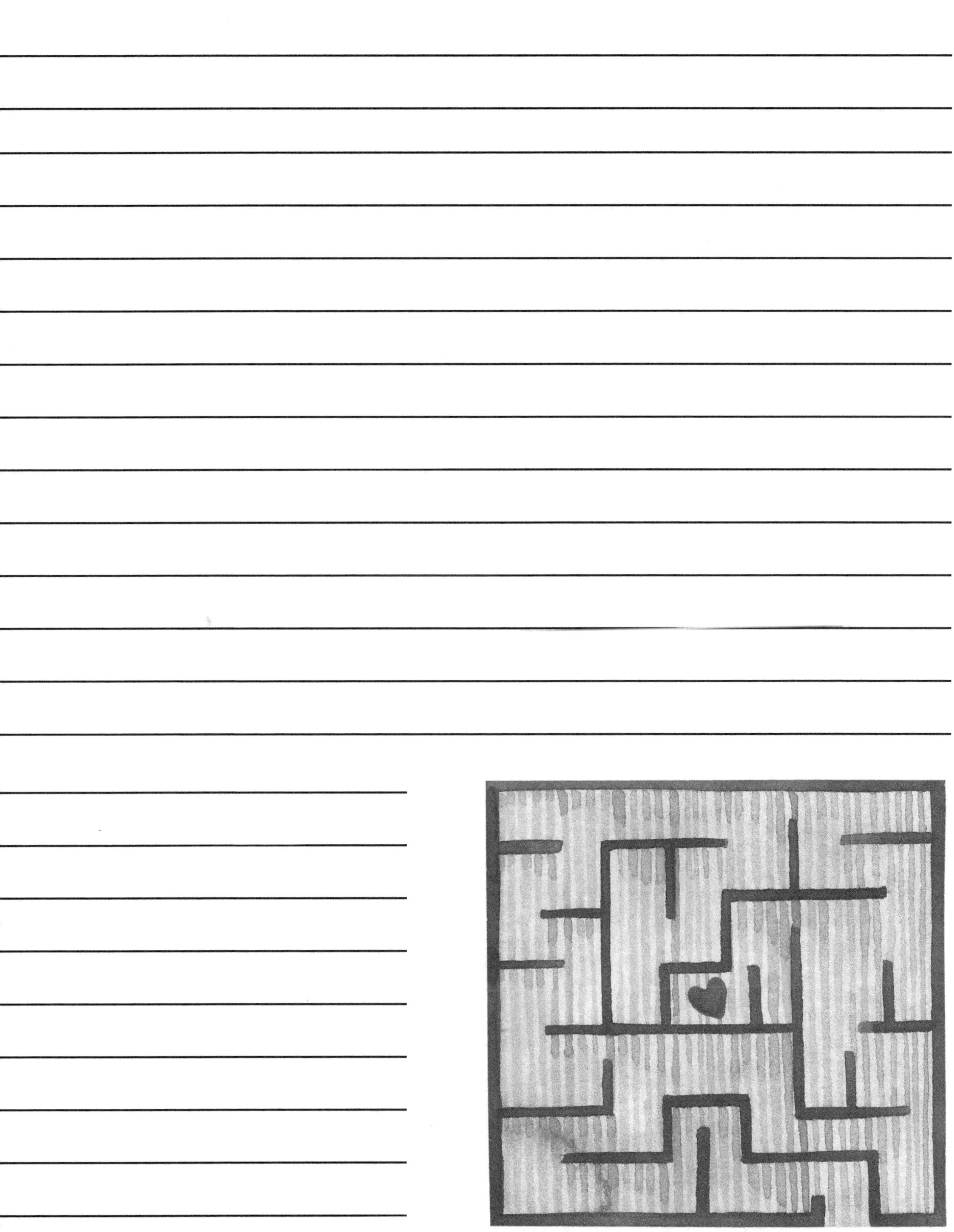

- Acts of Kindness Tracker -

- Acts of Kindness Tracker -

- Acts of Kindness Tracker -

Notes

Notes

Notes